INTERIM SITE

F

When Is Now?

When Is Now?

Experiments with Time and Timekeeping Devices

by HENRY HUMPHREY and
DEIRDRE O'MEARA-HUMPHREY

DOUBLEDAY & COMPANY, INC.
GARDEN CITY, NEW YORK

This book is for our daughter, Nora.
And it's about time.

LIBRARY OF CONGRESS CATALOGING IN PUBLICATION DATA
HUMPHREY, HENRY, 1930–   WHEN IS NOW? BIBLIOGRAPHY:
P. 79 SUMMARY: GIVES DIRECTIONS FOR MAKING SEVERAL
TIME-KEEPING DEVICES SUCH AS A SUNDIAL AND WATER CLOCK
AND DISCUSSES SOME ANCIENT KNOWLEDGE ABOUT MEASURING
TIME. 1. TIME—JUVENILE LITERATURE. 2. TIME
MEASUREMENTS—JUVENILE LITERATURE. [1. TIME. 2. TIME
MEASUREMENTS] I. O'MEARA-HUMPHREY, DEIRDRE, JOINT
AUTHOR. II. TITLE. QB209.5.H85 529.1
LIBRARY OF CONGRESS CATALOG CARD NUMBER 77–15158
ISBN: 0-385-13215-8 TRADE
    0-385-13216-6 PREBOUND

CONTENTS

PREFACE   7

CHAPTER I.   IT'S STILL EARLY   9

CHAPTER II.   WHERE? UNDER THE SUN   20

CHAPTER III. WHAT'S NEW UNDER
              THE SUN?   35

CHAPTER IV.   NIGHTTIME   56

CHAPTER V.   IS THERE A NOW?   69

        APPENDIX   77

        BIBLIOGRAPHY   79

PREFACE

TIME WAS no more important to primitive people than it is to a newborn baby. The first men just hung around and hoped that something to eat would either walk or grow nearby. That doesn't mean that they just lay around in their caves all day. They would hunt in the neighborhood, but they didn't start migrating with herds of animals until much later. There has to be a *need* for time for the *idea* of time to be important. Like prehistoric people, an infant just stays in one place and screams if food doesn't show up. But, as the baby grows, there is an ever-increasing need to understand what time means.

When babies grow up and become your age, time really matters. You have to know when to get up. The only way you can tell when to get up is by figuring out how long it takes you to get dressed and get to school. You've done this so often that you don't even have to think about it anymore. But when you started going to school all this had to be worked out.

That's an important thing to realize about time. It's

become such a part of our daily routine that we just don't give it much thought.

Try an experiment. Ask somebody if they know what time is. They'll probably say something like, "Sure, I know." Then ask them, "O.K., what is it?"

The answer to that question is not so easy.

Chapter I

IT'S STILL EARLY

LOOK UP the word "time" in the dictionary. You'll find that it's a noun, but we know it's not a thing or a place. And yet, if you ask people if they know what time is, they will most likely tell you that they do and they will believe that they do. They may explain it by saying something like, "Well, it's the space between, say, one and two o'clock." Then ask them to define it without referring to a clock or other time-measuring device. If a clock must be part of the definition, it's like defining the word "gasoline" by referring to the gas gauge in a car. Similarly, if you ask someone to tell you what gasoline is, they will probably reply, "Gasoline is what makes the gauge in our car go from 'empty' to 'full.'" That's like "the space between one o'clock and two o'clock." Gasoline is many things besides the space between "empty" and "full," just as time is many things besides the space between numbers on a clock.

Without mentioning "clock" or some watchlike gadget, could you describe time? It's not easy. Perhaps we can get some clues by following human progress in developing ways to think about and measure time.

Early people noticed that the sun seemed to move through the sky. Therefore, shadows that were long at dawn gradually shortened as the sun moved closer to its midday position (often called "high noon" because then the sun is as high in the sky as it gets that day). The shadows lengthened again as the day wore on. The next sunny day, see if you can discover this for yourself. Put a straight stick in the ground and every so often put a pebble or some other marker (one that won't blow away) at the far end of the stick's shadow. At the end of the day, connect the markers with a piece of string. Does the string make a straight line or does it make a curve? If the sun were up 24 hours a day, as it is near the north and south poles whenever it is "summer" there, the sun's shadow would make a complete circle. But in most places the summer/winter shadow patterns and the spring/fall shadow pattern will look like this.

As people began to band together, they had to work out some kind of system for keeping track of events. Before there was industry, smoke, and city lights, the night skies were often pretty clear. There wasn't any TV to watch or even any books to read, so people looked at the stars a lot. At first it looked pretty mysterious. But as the earth revolved, certain stars seemed to come and go at particular seasons of the year. Of course, the sun rose in the morning, climbed overhead, and set in the evening. Then there was the moon, which, then as now, appeared as a new moon once a month. (In fact, it could have been "once a moonth," but at some point one of the "o's" was dropped.) By the way, a moon month is usually called a "lunar" month.

Put the sun, the stars, and the moon together and it begins to look like the start of a timekeeping system. It takes the moon just about one month to go from one new moon to the next; it takes a year for every star (except the North Star, also called polestar or Polaris, which never seems to move at all) to seem to go around the earth and return to its previous position.

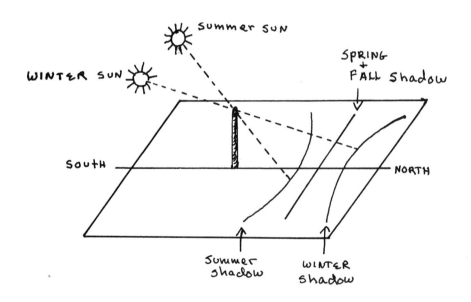

Paths of the Sun's Shadow

Star Trails Around the Polestar

Since the stars change positions during each of the four seasons—even during each succeeding night—all that the people had to do was look up and there was a sky calendar and clock of sorts. It was centuries before all this was organized into calendars and even more centuries before mechanical clocks were invented.

The astronomical knowledge and timekeeping system of ancient Mesopotamia is just now being slowly figured out from clay tablets that were unearthed some time ago. From these we have discovered that Mesopotamians had a calendar with a week of 7 days. They divided the whole day of 24 hours into 12 periods. Each of these periods was equal to 2 hours. They divided these into 30 parts. One of them was equal to 4 of today's minutes. Alas, this system was lost to the world until very recently. If only there had been printing presses, the Mesopotamians could have published their discoveries and we wouldn't have had to start all over again.

In ancient cultures ordinary people had very little knowledge of time except that the sun rose and set and the seasons changed. A knowledge of more accurate time was reserved for priests and royalty. It was only in the eighteenth century that everyone had their own timepiece or could read a public clock.

As ancient men watched the sky, they imagined that the brightest lights in the heavens were their gods. So it's not too surprising to discover that the names of the days of the week came from the names of the most luminous bodies in the sky. But why should there be 7 days in a week? There were probably three reasons for it:

1. Seven was a lucky number, even long ago.
2. It's roughly one quarter of a lunar month. Thus, there would be 4 weeks, each consisting of 7 days, to a month.
3. Genesis of the Old Testament says that God created Heaven and earth and everything on it in 6 days. On the seventh he rested.

There is a great coincidence in the fact that each day's name comes from the same heavenly objects, in the same order, in a great variety of languages, as shown in the following chart.

DAYS OF WEEK IN DIFFERENT LANGUAGES

ORIGIN	SAXON	SPANISH	LATIN
Sun	Sun's Day	Domingo	Dies Solis
Moon	Moon's Day	Lunes	Dies Lunae
Mars	Tiw's Day	Martes	Dies Martis
Mercury	Woden's Day	Miércoles	Dies Mercurii
Jupiter	Thor's Day	Jueves	Dies Jovis
Venus	Frigg's Day	Viernes	Dies Veneris
Saturn	Saterne's Day	Sábado	Dies Saturni

ORIGIN	SANSKRIT	BABYLONIAN
Sun	Ravi vara	Shamash
Moon	Soma vara	Sin
Mars	Mangala vara	Nergal
Mercury	Budha vara	Nabu
Jupiter	Brihaspati vara	Marduk
Venus	Shukra vara	Ishtar
Saturn	Sanischara vara	Ninib

Days of the week weren't too difficult to determine, but people didn't agree about the number of months in a year. A moon month is about 29 days. Early man, including the Egyptians, used the cycle from new moon to new moon for their month. Even today the lunar month is still the basis of the Buddhist calendar. The moon is not attached to the earth in any way; it revolves around the earth at a different speed than the speed at which the earth spins. So the actual number of days from new moon to new moon is not exactly 28 days. It's actually 29.53059 days. That's an awkward fraction for slightly more than 29½ days.

What the Egyptians did was to round it off to 30 days. So they had 12 months of 30 days each—or 360 days in a year. But they were smarter than that. The Egyptians had observed that there were a little more than 360 days between the longest day of one year to the longest day of the next year.

The sun year is usually called a "solar" year. The Egyptians' lunar year was running 5 days short of the solar year. To make up this difference, they added 5 days to the end of the year. Since these 5 "extra" days were not assigned to any month, they made the 5 days a holiday and a time for feasting and having a good time. The Egyptian calendar came close to being correct, but after 100 years or so it began to be out of step with the sun. It took centuries to develop a calendar that accurately followed the sun's seasonal progress.

What is the purpose of a calendar? Well, if you have

a calendar, you can tell when important days are coming up, such as your birthday, Christmas, or when summer vacation begins. For a calendar to be useful, it must show Christmas as coming just after the shortest day of the year. You know what the shortest day of the year is; that's when the sun rises at its latest in the morning and sets at its earliest in the afternoon. The shortest day of the year is, of course, the first day of winter. Now, if the calendar was not following the solar year exactly, it would probably be years before anybody would notice it. But after about 100 years people might begin to notice that Christmas is beginning to fall *before* the shortest day of the year. And if nothing was done to correct the calendar, in another 100 years Christmas might fall on a day in late autumn. When a calendar falls out of step with the solar year, it's wrong or in "error."

To avoid something like this, the calendar was constantly being fiddled with and changed. After the Egyptians, the Roman emperors tried to make the calendar more precise. When the emperors became less powerful and the Catholic church became more important, the leaders of the church became responsible for keeping the calendars in time with the seasons.

The last and most interesting change was accomplished in 1582 by order of Pope Gregory XIII. The changes were made after leading astronomers submitted numerous calculations. The solar year was determined to be 365.2422 days long and not the 365 days of the Egyptian calendar. In order to remove the error that

Moon

had crept into the existing calendar, the pope ordered that the day following Thursday, October 4, 1582, would be Friday, October 15, 1582. He also made another important change. On the old calendar the first day of the new year had been March 25. He changed it to January 1.

This most recent and most accurate calendar is the one we use today, but it is still not perfect. Its degree of error is 26 seconds per year, which means that it will be off a whole day every 3,323 years. So it'll be a long time before this calendar will get out of step with the seasons.

Even though the rest of Europe changed to the Gregorian calendar in 1582, England did not and continued to use the calendar first used when Augustus was Emperor of Rome in the year 8 B.C. When Britain and the colonies finally changed over to the Gregorian calendar in 1752, there was already an error factor of 11 days. There were riots in England when the change in calendars was made. One day it was October 4, 1752, and the next it was October 15. The riots were caused by workers who felt they had lost 11 days' pay. What was hardly ever mentioned during those disturbances was they had also lost 11 days' work.

The following chart compares the accuracy of some of the best known Roman calendars:

DEVELOPMENT OF THE ROMAN CALENDAR

ROMULUS 738 B.C.	NUMA 713 B.C.	COUNCIL OF DECEMVIRS 451 B.C.	JULIUS 47 B.C.	AUGUSTUS 8 B.C.	GREGORY XIII 1582* 1752**
Martius 31	Januarius 29	Januarius 29	Januarius 31	Januarius 31	January 31
Aprilis 30	Martius 31	Februarius 28	Februarius 29–30	Februarius 28–29	February 28–29
Maius 31	Aprilis 29	Martius 31	Martius 31	Martius 31	March 31
Junius 30	Maius 31	Aprilis 29	Aprilis 30	Aprilis 30	April 30
Quintilis 31	Junius 29	Maius 31	Maius 31	Maius 31	May 31
Sextilis 30	Quintilis 31	Junius 29	Junius 30	Junius 30	June 30
Septembris 31	Sextilis 29	Quintilis 31	Julius 31	Julius 31	July 31
Octobris 30	Septembris 29	Sextilis 29	Sextilis 30	Augustus 31	August 31
Novembris 31	Octobris 31	Septembris 29	Septembris 31	Septembris 30	September 30
Decembris 29	Novembris 29	Octobris 31	Octobris 30	Octobris 31	October 31
	Decembris 29	Novembris 29	Novembris 31	Novembris 30	November 30
	Februarius 28	Decembris 29	Decembris 30	Decembris 31	December 31
304 days	355 days	355 days	365¼ days	365¼ days	365.2422 days

* EUROPE
** ENGLAND

Source: Harrison J. Cowan, *Time and Its Measurement from the Stone Age to the Nuclear Age.*

Chapter II

WHERE? UNDER THE SUN

BECAUSE OF ITS favorable climate, humans settled not too far from the Mediterranean Sea (which in Latin, the language of the Romans, means "middle of the earth" —for it *was* the center of their world) and alongside the fertile river valleys of the Tigris, Euphrates, and the Nile.

The weather was so dry on the edge of the desert, where the Egyptian kings, the Pharaohs, had their tombs built, that it made it possible to keep the mummies (the kings' preserved bodies) in good condition as well as all sorts of arts and crafts that were put in the tombs to comfort the dead kings during their long journey to rebirth. Not too long ago scientists started discovering all these things in the tombs. That's how we found out so much about the way they lived and about their system for measuring time.

For all these reasons, Egypt makes an ideal starting place to find out how people got from the understanding of years, weeks, and days to understanding and measuring hours. As a matter of fact, about 10,000 years before the birth of Christ, the Egyptians figured out a device that would keep track of hours. Wow! Hours! That split time into small enough pieces for all their needs. Even the first mechanical clocks, which didn't begin to appear until almost 12,000 years later, measured only hours.

"Hours" are really an idea mankind created for its convenience. Hours are not a scientific fact. Without any mechanical help, it's easy to see that the seasons change, that the time between new moons is a month, that from sunrise to sunset is a day, and that 7 of these can be called a week. But to divide the day into smaller parts is a lot more difficult. With this in mind, it's all the more extraordinary that the Egyptians, thousands of years ago, invented a clock for dividing the day into hours.

The Egyptians developed a very practical math to help them carry on business and all the building (including the pyramids). They even invented a basic geometry so that they could, among other things, redraw the boundaries of their farms along the Nile after the annual flood. The river overflowed its banks every year. It was not a tragedy but a blessing. After the flood waters returned to the banks of the Nile, a new coating of silt, or rich earth, carried from lands near the source of the river would be dropped on the farmlands.

At any rate, the Egyptians divided each day into 24 hours. There were 12 "night" hours that were marked by priests observing the changing positions of the stars.

There were 10 "daylight" hours and 2 "twilight" hours, one for dawn and another for dusk.

Here's how you can make a copy of the earliest known sun-clock (or "time stick," as the Egyptians called it), first made about 10,000 years before the birth of Christ.

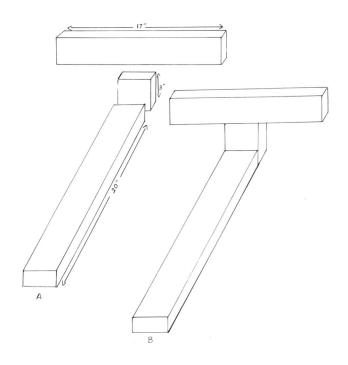

How to Make an Egyptian Time Stick

To make an Egyptian time stick, you will need a length of 1"×2" lumber 40" long, light-colored paint, a small amount of black paint (for hieroglyphs), and ½" nails.

Saw the 1"×2" into 3 sections measuring 3", 17", and 20". Nail or glue the sections together as in drawing A. When the time stick is assembled, it will look like diagram B.

Egyptian Time Stick

Paint the time stick a light color so that the shadow will show up clearly.

Place the stick outdoors at seven o'clock in the morning on a sunny day. Use a compass and turn the stick so that the back of the top bar is facing east. Make sure that the stick is level. You can use a carpenter's level if you have one, or put a glass bowl half-filled with water on top of the stick. The water in the bowl will be level even if the bowl is not. Compare the water level with the stick. Adjust the stick so that it is as level as the water.

As soon as the time stick is in the correct position, make a light pencil mark where the shadow falls. At eight o'clock make another mark where the shadow falls, and so on until eleven o'clock. These marks are

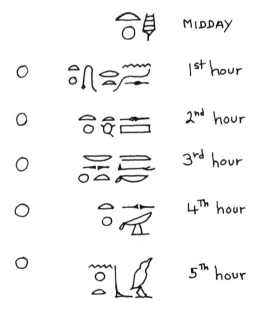

MIDDAY

1st hour

2nd hour

3rd hour

4th hour

5th hour

Egyptian Hieroglyphs

the 5 hours before noon. Leave the stick until noon. At noon, turn the time stick, in place, facing west, again making sure that it's level. Now the shadow, as it progresses, will fall on the same markings but in reverse order. These are the 5 hours after noon. There is a hieroglyph (a single character in Egyptian picture writing) for noon, but it is not counted as one of the 5 hours. The sun is overhead at noon and does not cast a significant shadow.

After you have marked the hours, you can paint the Egyptian hieroglyphs for each hour on the stick.

Don't expect your model of the Egyptian time stick to give you the exact time shown on your modern clock. It doesn't for the same reasons that sundials, which were invented thousands of years after the Egyptian time stick, don't agree with your clock or watch. The main reasons why sun-clocks don't usually agree with watches are explained in the next chapter.

As civilization began moving northward from sunny Egypt to cloudy Europe, the limitations of the sun clock were pretty obvious. As people moved north, they took another Egyptian invention with them that was almost completely independent of the sky. It did not need the sun to operate; it needed water. It is called the "clepsydra," a Greek word meaning "water thief" because it seemed to make water disappear. It worked by measuring how fast water in a top container poured through a small hole into a bottom container. Here's how to make a simple version of the clepsydra:

Egyptian Water Clock

How to Make an Egyptian Water Clock

MATERIALS

1 piece of Plexiglas tubing ⅛" thick, 4" in diameter, and 7" tall (This is for the lower part of the clock.)

1 piece Plexiglas tubing ⅛" thick, 4" in diameter, and 6" tall (This is for the upper part of the clock.) If the measurements aren't absolutely exact, it doesn't matter. The important thing is that each tube should hold about a quart of water.

2 pieces of Plexiglas ⅛" thick and 4½" square (These are for the bottom of the tubes.)

1 piece of Plexiglas ¼" thick, 24" long, and 4½" wide (It can be clear or any color you like.)

1 small can of acrylic solvent cement (Epoxy might do, but solvent cement is better.)

ASSEMBLY

1. Sand the edges of the tubes until they are flat and smooth. Lay the sandpaper flat on a table. Hold it down with one hand. Hold one open end of the tube against the sandpaper and sand with a circular motion. Sand the other end and repeat with the other tube.

2. You can either cement the 2 tubes onto 2 square pieces of Plexiglas or you can cut the squares into

round pieces to fit against the bottom of the tubes. You'll need a jigsaw if you want to cut the round pieces. Leave the masking paper on the Plexiglas while you are sawing.

3. To cement the tubes to the bottom pieces, do this: Pour a small amount of solvent cement into a saucer, about $\frac{1}{16}''$ deep. Dip the bottom of the tube into the saucer. Hold it there for half a minute or until the bottom of the tube is a little soft and does not click against the saucer when you tap it. Place the tube onto the bottom piece right away. Hold it for a minute until it sets. Do the same with the other tube.

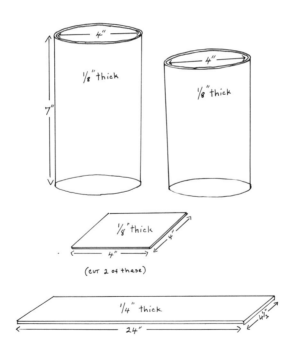

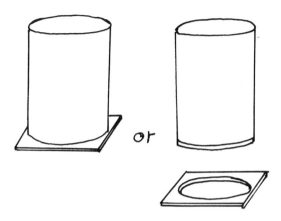

4. Fill one tube with water. If the bottom leaks, apply more solvent cement where the tube and the bottom join. Use an eyedropper or a fine paintbrush. Do the same with the second tube.

5. Drill a hole in the bottom of the short tube. Use the finest drill bit possible, such as a bit for a model maker's pin vise. If it breaks, use one a little heavier.

6. Smooth the edges of the long (4½″×24″) piece of plastic with sandpaper. Drill a 1″ diameter hole in the middle.

7. Measure 9½″ in from each end and make 2 marks, as shown in the diagram. Use a heating strip for bending plastic or have an adult use a propane torch to heat the plastic at one of the marked places. When it is soft, bend it to form a right angle. Do the same at the other marked place.

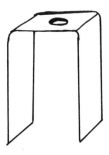

8. Place the tall tube underneath the stand. Fill the short tube with water and place it on top of the stand. Start timing the water as it fills the bottom tube. After an hour make a mark where the water level is. If the water is dripping too fast and the bottom tube is filling up too fast, plug the drip hole with soft modeling clay. Push a pin through it to make a smaller hole. Time the dripping for 3 or 4 hours. The spaces between the hour marks should be equal, so you can just mark all the way to 12 without any more timing.

9. You don't have to start numbering the hours at one o'clock in the morning. You might like to start off your clock when you get up in the morning, around seven o'clock perhaps. Then the water will not reach the top numbers, and will not have to be refilled, until about seven o'clock at night.

10. You can paint the numbers on your clock or you can use LETRASET®, which is made up of sheets of letters and numbers that can be pressed on with your fingers. It is available in most art-supply stores. (We used 30-point LETRASET® roman numerals.) LETRASET® is quite expensive. The band at the top and bottom of the short tube was painted on (terracotta color). The "Greek key" design was a sticky tape border that was put over it. The design came in a whole sheet of borders, somewhat like LETRASET® but not as expensive. This particular Greek key design is called FORMATT® ※6420.

An Alternate Way to Make
the Egyptian Water Clock
Out of a Coffee Can and a Jar

MATERIALS

1 quart-sized tin can, such as a 2-pound coffee can
2 wooden boards 9½" long and 4½" wide
3 strips of wood 5" long and about 1½" wide
1 quart glass jar such as for mayonnaise

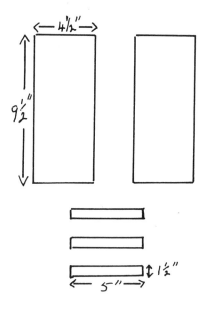

ASSEMBLY

1. Paint the inside and outside of the can with acrylic paint. This will help keep it from rusting and will also make it look nice. You can decorate the outside if you want to.

2. Using a needle (if possible) and a hammer, tap a *very tiny* hole through the bottom of the can.

3. Construct the stand out of the wood as shown in the diagram. Nail 2 of the strips across the top of the upright boards and 1 strip across the back.

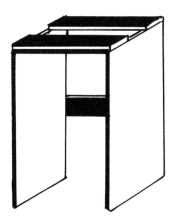

4. Put the jar underneath the stand, fill the can with water, and set it on top of the stand. If the water drips too fast, cover the hole with masking tape or electrical tape and make another smaller hole.

5. Time and mark the jar the same way as in step ✳8 in the plastic water clock instructions.

Much more complicated water clocks were made later on and they were surprisingly accurate. Water would flow slowly into the bottom container and a float, with a gear mounted on it, would rise and engage another gear, which in turn, moved the pointer along a cylinder marked with hours and, in newer clocks, even minutes. These clever clocks, with their gears, led to the invention of purely mechanical clocks.

Chapter III

WHAT'S NEW UNDER THE SUN?

WHILE A LOT of work was going on in attempts to perfect the water clock, people were still trying to make a sun-clock that was easier to use and more accurate than the old Egyptian time stick. You remember that one— it faced east in the morning and had to be moved to face west in the afternoon.

Long after the Egyptian sun-clocks had been invented, sundials were designed that did not have to be moved morning and noon. The important difference between sun-clocks and sundials was the pointer, or "gnomon." (The word comes from a Greek word meaning "one who knows." The "g" is not pronounced.) The earliest sundials had a gnomon that was perpendicular to the dial and time was told by measuring the length of the shadow. A later model replaced the vertical pointer with one pointing to the polestar (which is another way of saying that the gnomon was set to the angle equal to the latitude of the place where the sundial was located). The time was now told by the direction of the shadow and not by its length; this meant that greater accuracy was possible. Here's how to make one.

How to Make a Sundial

LAYOUT

In making a sundial, the first step is to lay the design out on paper. Since it's a fairly complicated process, the following instructions have been made as clear as possible. The layout diagram should be drawn on a piece of paper about twice the size of the finished dial. Step ✳5 below clearly tells you how to make your sundial the size you want it to be. Basically, the size of the dial depends on how long you make the hour lines. The hour lines on a sundial are laid out according to the degree of latitude where it will be placed. Latitude is shown on maps as lines going east–west. If you plan to put your sundial in the garden, you must find the latitude of your town or the nearest city. The chart in the Appendix will be of some help. Or you can contact a land surveyor in your area who might have this information. The diagram, as drawn, is for latitude 41°. If you are lucky enough to live in that latitude, or a degree or two on either side, you can trace the lines directly from the diagram. Since the diagram in this book is smaller than the size of the dial face you will be making, start from point 0 and ex-

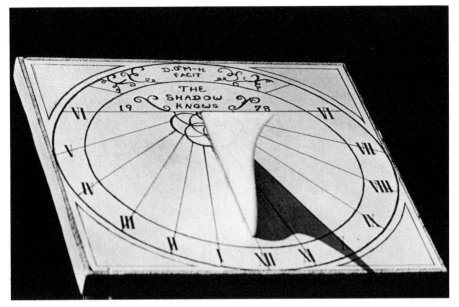

Sundial

tend the lines shown to the size of your dial. Mark each line with the appropriate hour. Then you can add the "furniture." Dial "furniture" consists of information and decorations that do not actually tell the time of day. In other words, everything except the radiating hour lines and their numbers is considered sundial furniture.

If you live away from 41° latitude, you must draw your own layout. Although it looks complicated, if you follow every step very carefully, making a check mark on each of the following steps as they are finished, you will eventually wind up with this wonderful scientific-looking drawing.

37

ASSEMBLY

After you have finished the layout diagram, we'll tell you how to finish the sundial. Here we go.

1. Find the degree of latitude where you live. (*You will need a ruler and a protractor.*)

2. Draw line AB (On the finished dial this will be the six o'clock line.) Mark point O at the center of line AB.

3. Draw line CO at a right angle (90°) to line AB. This will be the twelve o'clock line.

4. Draw line OD. The angle COD should be equal in degrees to your latitude. Use a protractor. Place the protractor so that the 0° mark is on line CO and the 90° is on line AB. Make a mark where the degree on your protractor is the same as your degree of latitude.

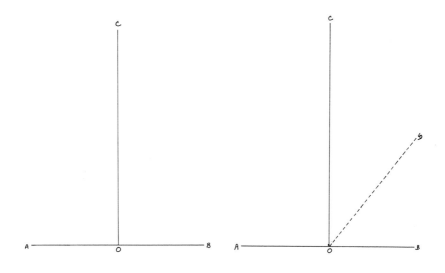

5. Somewhere along line OD make mark E. If you put E close to O, your whole diagram will be quite small; if you put it fairly far from O, the diagram will be larger. Here's an example. If you place the E mark 1½″ from O, the whole diagram will be about 2½″×3″. If you place it 7″ from O, the whole diagram will be 12″×15″. Knowing this will help you to draw the correct size diagram you want.

6. From mark E, draw a line at a right angle to OD and extend it as far as line OC. Where it meets OC make mark F at this point.

7. On line OC make mark G. It should be the same distance from point F as F is from E.

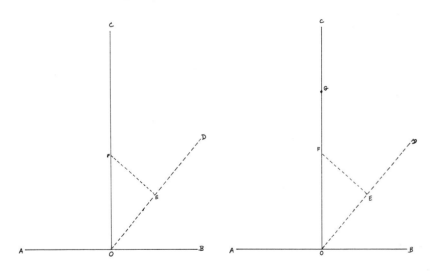

8. Through F draw line QR. Make it parallel to AB.

9. Through G draw line ST. Make it parallel to AB and QR.

10. Using a protractor, draw a semicircle with point G as its center; it should begin and end on line ST. The semicircle can be any size as long as it doesn't extend below line QR. Keep the protractor in place for the next step.

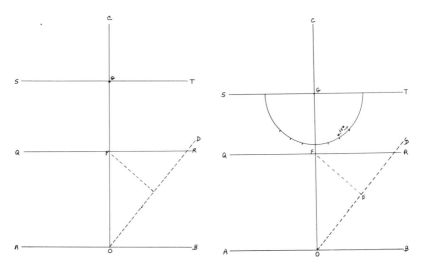

11. Now make 3 marks, 15° apart, to the right of line GF. Then make 3 marks, 15° apart, to the left of line GF. In the diagram the marks are on the arc of the semicircle.

12. From point G draw lines through each of the 15° marks and extend them to line QR. Mark the 6 lines, plus the F point, with the hours 9, 10, 11, 12, 1, 2, 3, as in the diagram.

13. From point O draw connecting lines to 9, 10, 11 and 1, 2, 3. The twelve o'clock line is already drawn in —it's the OF line. These will be the hour lines for these numbers on the finished dial.

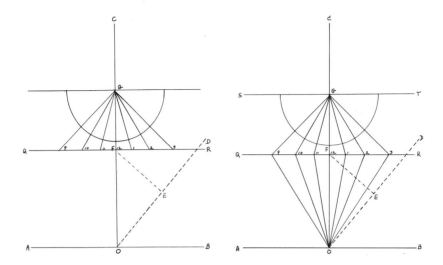

14. Draw line FW parallel to the nine o'clock line (the line extending from 9 to point O). Mark the W point on line AB.

15. Where line FW crosses the one, two and three o'clock lines, mark K, L, and M.

16. On line FW you now have to mark N and point P. The distance from point M to point N should be the same as the distance between points L and M. Measure and mark point N. Now mark point P. The distance from point M to point P should be the same as the distance between points K and M. Measure and mark point P.

17. Draw line YV from the nine o'clock point down to line AB. It should be parallel to line FO. Draw line XW from the three o'clock point down to line AB. It should also be parallel to line FO.

18. From point O draw a line through point N (on line FW) all the way to line XW. Do the same through point P (also on line FW). These are the four and five o'clock lines. Mark them 4 and 5. You're now in the home stretch!

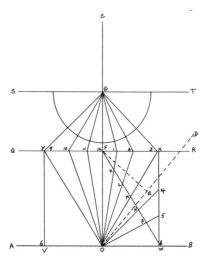

19. On line YV mark the points for seven and eight o'clock. The distance between Y and 8 should be the same as that between X and 4. Measure and mark the point for eight o'clock. The distance between Y and 7 should be the same as that between X and 5. Measure and mark the point for seven o'clock. Now draw a line from O to 7 and a line from O to 8.

From point O the numbers read clockwise from 6 A.M. to 6 P.M. If you would like to include 5 A.M. and 7 P.M., it's easy to do. For 5 A.M. just extend the 5 P.M. line through point O all the way over to the A.M. side. For 7 P.M. extend the 7 A.M. line through point O to the P.M. side. If you put a straightedge along the 5 P.M. line, you will see where the 5 A.M. extension falls; you can do the same with the 7 A.M. line.

Now that you have laid out the hour lines, you still have to make a design for the completed dial face.

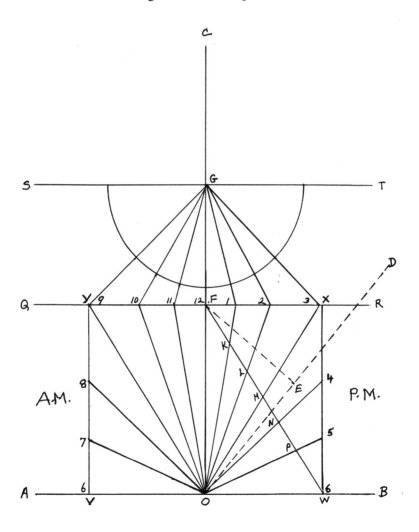

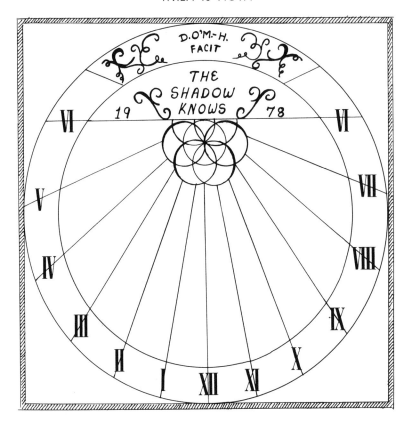

1. With a drawing compass, draw a 10⅛″ diameter circle. One inch in draw another smaller circle. Mark the center with a pencil dot.

2. Using a ruler, measure 2½″ up from the center dot and make a mark. Draw a line through this mark from one edge of the outer circle to the other edge of the outer circle. This horizontal line becomes the six o'clock line. Make a mark exactly in the middle of the six o'clock line. This is the O point from which all of the hour lines radiate.

3. Match the O point in your original diagram with the O point in the circle. Make sure the six o'clock lines match. It doesn't matter if the diagram is larger or smaller than the circle. No matter how long you make the lines, the O point and the angles will always remain the same. Tape the diagram down over the circle so it won't move.

4. Slip carbon paper between the diagram and the drawn circle. Trace *only* the hour lines from point O within square VWXY. Do not trace lines OD, FE, or FW.

5. Using a pencil, mark each hour line with the proper number: 6, 7, 8, and so forth. This is just for reference; you don't need to put in the roman numerals until you transfer the finished design to the wooden dial face.

6. Draw a square border around the outside of the circle. Add any "dial furniture" you may feel up to: the name of the marker (D. O'M.-H. are my initials and *fecit* means "made by" in Latin), the date, curlicues, or an appropriate motto. Here are some sample mottos:

"I count only the sunny hours."

Tempus fugit
   ("time flies" in Latin)

Serene I stand
Amidst the flowers
To tell the passing
Of the hours.

Sine sole sileo
   ("Unless the sun shines I am silent" in Latin)

45

Dial furniture can get quite complicated. You can borrow books from the library that tell all about sundials and show many different kinds. Some of these books are listed in the Bibliography at the back of this book.

7. There is one more thing to do before you make the actual sundial. You have to lay out the gnomon (pointer). The gnomon is the same angle and approximately the same shape as lines FOE in the diagram. Use the picture as a guide, but remember to make the FOE angle the same degree as your own latitude. OF is the base line and it's placed along the twelve o'clock line from point O to the twelve o'clock point. The gnomon can be a simple triangle or it can be made into a gracefully shaped pointer. You must add a tongue to the base of the gnomon so that it will fit into a slot in the dial face and hold the gnomon perpendicular. Base line FO should be as long as the distance from O to the twelve o'clock point on the dial face.

HOW TO COMPLETE THE SUNDIAL

1. Cut a piece of ½" thick plywood or shelving 10½" square.

2. Paint the square white with oil-based or acrylic paint. Let it dry thoroughly.

3. Transfer the dial design to the painted wooden square with carbon paper. Tape the carbon paper and the drawing to the wood so it won't slip while you're tracing. After the design is transferred, remove the tape, the carbon paper, and the drawing. Go over the traced lines on the wooden square using either black paint and a fine brush, a pen and India ink, or a perma-

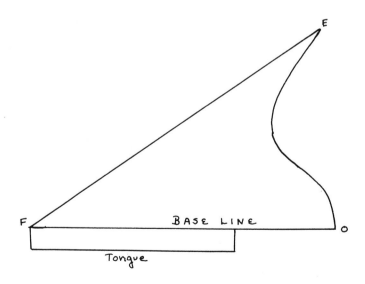

nent felt marker. Be careful about using a felt marker; the varnish that you spray on to protect the finished drawing will often make the ink from the marker run. Add the hour numbers at this point. If the roman numerals in the photo look extra neat and professional, it's because I used LETRASET® numbers. Ask for LETRASET® roman numerals #3226.

4. Spray the sundial with a fixative (spray varnish). Give it four or five thin coats. If you used a felt marker, spray very lightly and don't move the dial until it's dry.

5. Using an ½″ drill bit, drill a hole at the O point. With a jigsaw or saber saw make a slot large enough for the tongue of the gnomon to fit into.

6. Cut the gnomon out of a ⅛″ thick piece of wood, thin metal, or ⅛″ or thinner plastic. If you use clear plastic, paint it opaque. Fit the tongue of the gnomon into the slot of the dial face. Now you have a sundial!

Mount your sundial on a pedestal in a spot that is sunny all day long. The position of the sundial is very important. The twelve o'clock line must face due north. The dial face must also be level. Use a carpenter's level to make sure that it is. If it is correctly positioned, you will find that at night the gnomon points directly at the polestar.

This later model sundial, which is similar to the ones found in gardens today, reflected the cleverness of Arab mathematicians and astronomers of the Middle Ages. A sundial of this type appears to have been quite rare in

Europe before the end of the 1400s. Mechanical clocks had already been developed by that time. But many years were to pass before the new clocks ran with any real accuracy. So the sundial was often preferred by people who wanted a more accurate time than the early mechanical clocks could provide.

A Hand Sundial

You may remember that until fairly recently only priests and royalty could know the correct time. In the sixteenth century European peasants had a very primitive sundial that could give them a rough idea of the time. You can make one of these peasant's sundials with your own two hands. In fact, it's made out of your two hands. Hold a small stick close to the base of the thumb. Hold your hand horizontally, palm up, as in the photo. Hold the stick in your left hand pointed west for A.M. time and in your right hand pointed east for P.M. time. The stick held by the thumb should be tilted to an angle equal to the latitude of your location. Your teacher or your parents should be able to help you find out what that angle is in your area. You only have to find it out once—unless you and your family move to another town. The Appendix will give you an approximate angle for your area.

The positions of the shadows on your hands should be as follows for each of the daylight hours:

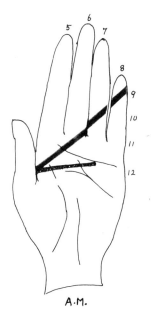

A.M.

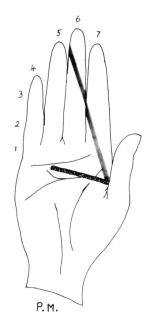

P.M.

One thing you should know about sundials is that they will agree with mechanical clocks only 4 times a year: around December 25, April 15, June 15, and September 1. On all other days the Egyptian time stick, sundials, hand dials, and other similar devices would be inaccurate to varying degrees. "Inaccurate" is probably too strong a word because a sundial always shows local apparent sun time accurately. This idea is kind of complicated, but we'll try to simplify it a bit.

Sundials or sun-clocks can only show "apparent solar time" (unless they are of a very involved design). This means that when the sundial shadow is shortest (or narrowest), the sun is at its highest point in the sky and the sundial shows twelve noon. When the sundial next shows noon, the sun will seemed to have made a complete trip around the earth and a full day will have passed. This is called an "apparent solar day."

The problem is that apparent solar days are not all the same length. In January the sun takes about 24 hours, plus half a minute by the clock, to run from noon to noon. In October it speeds up and makes the trip around the world almost a minute faster. For the purpose of legal standard time, all the variations are averaged out and, by international agreement, every day is understood to be exactly 24 hours long. Otherwise all the clocks and watches in the world would have to be adjusted practically every day. Anyhow, the legal 24-hour day is also called the "mean solar day." If it's really important to have your sundial or sun-clock check out with your watch, you can add or subtract minutes from your sundial readings by using the following chart:

Add or subtract the amount in the second column from sun time to get local mean time:

DATE	CORRECTIONS IN MINUTES
Jan. 1	+3
Jan. 8	+7
Jan. 15	+9
Jan. 22	+12
Jan. 29	+13
Feb. 3	+14
Feb. 26	+13
March 1	+12
March 15	+9
March 30	+5
April 4	+3
April 15	0
April 25	−2
May 1	−3
May 15	−4
June 4	−2
June 15	0
June 24	+2
June 30	+3
July 4	+4
July 10	+5
July 31	+6
Aug. 11	+5
Aug. 17	+4
Aug. 25	+2
Sept. 1	0
Sept. 10	−3

DATE	CORRECTIONS IN MINUTES
Sept. 30	−10
Oct. 6	−12
Oct. 15	−14
Oct. 26	−16
Nov. 22	−14
Dec. 1	−11
Dec. 9	−8
Dec. 17	−4
Dec. 25	0

There's another problem in comparing sun time to watch time. Suppose you are in the Eastern time zone, which stretches from Maine to Ohio. Even though the sun reaches the noon position in Maine minutes before it reaches the noon position in Ohio, all the clocks and watches will show the same time from Maine to Ohio. But that's not so hard to correct because all you have to do is find the difference between your local sun time and the standard time in your zone. You only have to find that difference once since you will always either add or subtract this difference from your sundial readings, along with whatever corrections are indicated on the list of dates. To help you determine the difference between your local sun time and local standard time, use the chart in the Appendix at the back of the book that shows the variations for quite a few of the leading cities in the United States.

But, honestly, we don't think you should knock yourself out trying to make some superaccurate timepiece out of a sundial or sun-clock. The fun part is to see how it works and to realize that, at least in the case of the Egyptian time stick, it's the way time was measured for more than 100 centuries.

Before we leave these sun-clocks and turn to other kinds, you should know something about the "astrolabe." The astrolabe is the most outstanding scientific instrument used in the Middle Ages. Invented by a Greek approximately 200 years before the birth of Christ, it was used to find the altitude of the sun or, at night, the altitude of certain stars from the horizon. By finding the altitude of these heavenly bodies, early navigators could locate the position of their ship in terms of degrees of latitude as well as the time of day or night, which would help them find their longitude. As you may know, on a map lines of longitude run north–south.

The astrolabe was used on board ship for centuries. (Sundials and water clocks could not be used because of the violent motion of ships caught in stormy seas.) In the seventeenth century, the astrolabe was replaced by the sextant, which is still in use today.

The astrolabe is a complicated device, so we won't try to show you how to make one. However, if you'd like to have one of your very own, there's a company that puts out a kit for making an astrolabe. The cost is approximately $20 and the astrolabe kit is available from Paul Macalister & Associates, Box 157, Arden Shore Road, Lake Bluff, Illinois 60044.

Chapter IV

NIGHTTIME

SUPPOSING YOU were a sailor and you needed to know the time of night when the sun was shining on a different part of the world. If you were living in the sixteenth century, you would use a nocturnal. A nocturnal is an instrument that makes it possible to tell time by means of the stars. The word "nocturnal" comes from the Latin and means "having to do with night."

How to Make a Nocturnal

Before you try to make your own nocturnal, examine the photo to see what a nocturnal looks like.

When in use, it is held up by the handle. The pole-star is sighted through the little hole in the center. The date is set, the pointer is aimed at one of the constellations, usually the Big Dipper and, lo and behold, the time of night is revealed!

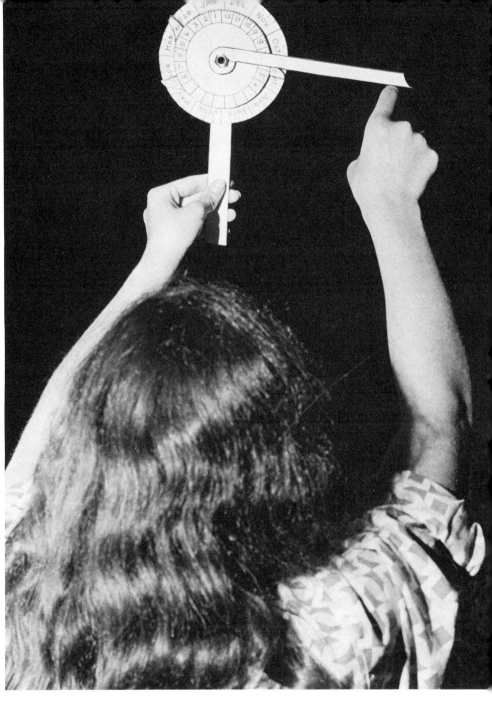

Nocturnal

MATERIAL

The nocturnal is made up of 3 pieces: a large disc showing the 12 months of the year; a smaller disc with several "teeth" that shows the 24 hours of the day; and a pointer, called a "cursor." The 3 pieces are held together in the center by means of a hollow fastener on which the discs and pointer can be turned.

You can make the nocturnal out of ⅛″ thick plywood, posterboard, sheet metal, or even acrylic sheet plastic. You'll need a protractor. You'll also need a threaded ½″ piece of pipe and 2 metal nuts. These are generally available at hardware stores and lamp-repair shops. After you've picked out the material you're going to use, here's what you do.

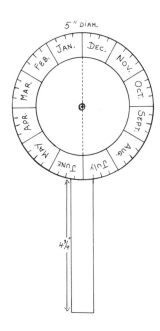

ASSEMBLY

1. Draw a circle on the cardboard or whatever material you're using. It should be about 4″ in diameter. If you don't have a drawing compass with which to make the circle, use a small bowl or a coffee can and trace around the edge.

2. Using a protractor, divide the circle into 12 even segments. Each segment will measure 30°. They are the months of the year. Divide each segment into 30 days if possible. If there isn't enough room to make 30 lines, then divide each segment into 4 weeks.

In one of the segments write "Jan." Then, continuing *counterclockwise* (from right to left), fill in the other months. You can abbreviate the long names so they will fit into the spaces. Mark the center of the circle with a dot. Draw a handle onto the circle opposite the January line. It will be cut out as part of the circle. If you forget to cut out the handle, it can be attached to the back later.

3. Draw a second circle about 3¾″ in diameter. Divide this circle into 24 segments with the protractor. They are for the 24 hours of the day. Pick one segment and label it "12M" (for midnight). The rest of the hours continue *counterclockwise* around the circle. Since stars are not visible during the day except very far north (in the wintertime), include only the night hours. That is why some of the segments have been left blank. If you live as far north as Alaska, for instance,

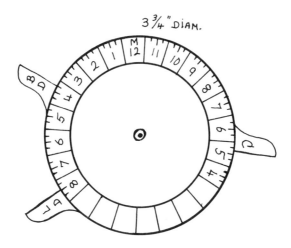

3 ¾" DIAM.

include any hours during which the stars are visible. Divide each of the hour segments used into quarter hours.

4. You will need at least 1 "tooth" on this 3¾" circle, but 2 or 3 are better. The 3 "teeth" represent 3 constellations from which you can calculate the time. Each tooth is marked with its proper constellation. The Big Dipper (the Romans called it "Ursa Major," which in Latin means the "Great Bear") is one of the most commonly used constellations for calculations of time. If, for some reason, the Big Dipper isn't visible (maybe clouds are hiding it), you have two other choices. Use the Little Dipper (Ursa Minor, or "Little Bear," to the

Romans) and Cassiopeia (named after an imaginary queen of Ethiopia). These are easily recognizable constellations since they are prominent in the skies of the northern hemisphere.

a. The straight, upright edge of tooth ⚹1 is at 63.4° *counterclockwise* from the midnight division. That will put the time at 4:14 A.M. Draw this tooth and mark it B.D. for Big Dipper.

b. Tooth ⚹2 is at 124.4° *counterclockwise* from 12M. This puts it at 8:18 A.M. Mark this tooth L.D. for Little Dipper.

c. Tooth ⚹3 is at 94.7° *clockwise* from 12M, or at 5:41 P.M. Label this tooth C for Cassiopeia. Be sure the teeth stay attached to the circle when you cut it out.

5. Now draw the pointer or cursor. Draw a small circle about 1⅛" in diameter. Draw the long side of the pointer so that it would pass through the center of the circle if it were extended back. The short side should be parallel to the long side.

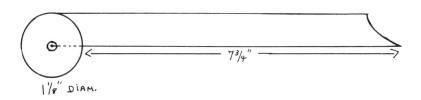

6. In order to assemble the nocturnal, you will need the short piece of threaded ½″ pipe and 2 nuts. Drill or otherwise make a hole through the center of each of the circles and the cursor. The hole should be just large enough so that the threaded pipe fits into it snugly. Put the 3 pieces together, the large circle on the bottom, then the small circle with the teeth, and the cursor on top. Secure them with the pipe and nuts screwed on. The nocturnal should be loose enough for the circles and the cursor to move easily, but not too loose.

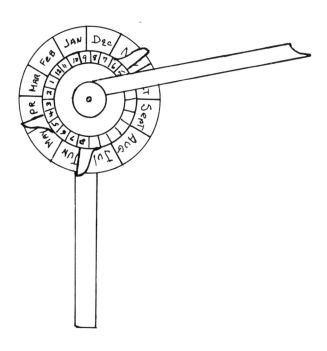

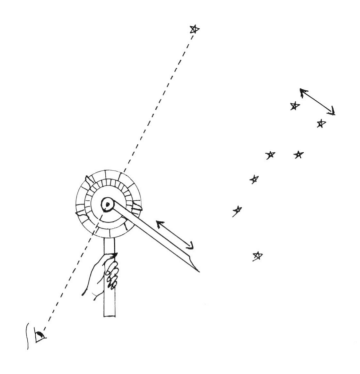

How to Use the Nocturnal

Choose the constellation that you will be using for your calculations. Then set the nocturnal at the correct date. If you are using the Big Dipper, set the tooth marked B.D. at the appropriate date. For example, if it's June 15, the tooth would be set at the middle of the June segment.

Hold the nocturnal in front of you at arm's length, with the handle pointing downward. Raise your arm and sight the polestar through the hole in the middle of the fastener that holds the pieces together. While keep-

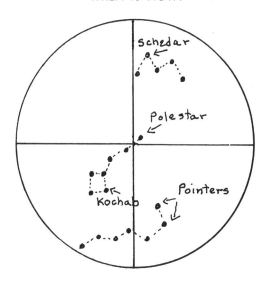

ing the polestar in sight, swing the cursor until the long edge lines up with the "pointer stars" of the Big Dipper. (The "pointer stars" are on the "dipper" part of the constellation; they lead the eye to the polestar.) The edge of the cursor will cut through the approximate time on the smaller disc. It will be correct to within 15 minutes.

If you are using Cassiopeia, line up the cursor with the star Schedar, as shown in the diagram. If you use the Little Dipper, line it up with Kochab, also shown in the diagram.

There were other methods of measuring time at night. These were better suited to countries where the skies are not as clear as they are in Egypt. One of these methods was the burning of a specially marked candle.

How to Make a Candle Clock

A candle clock must be just about the easiest kind of clock to make. You simply mark hour lines (or half-hour lines) from the top of a candle to the bottom. As the candle burns down to each line, it tells you the time.

First you will need to buy a candle that is not tapered at the top but straight up and down. The candle should not be more than 1½" in diameter. If the candle is too thick, it will burn down on the inside only and leave an outer shell. Beeswax burns the slowest and paraffin burns the fastest, so if you plan to stay up late get a slow-burning candle.

1. When you have chosen a candle, measure it from where it comes out of the candlestick holder to the top. If it's a 12" candle, it will probably measure 11½" (½" is in the candlestick holder).

2. Light the candle and let it burn for one hour (or half an hour if you decide to divide the candle into half-hour lines). After the hour is up, blow out the flame. Measure the candle again to see how many inches burned in an hour. Using that measurement, make a mark for every hour down the length of the candle.

To take an example, the candle in the photo burned 1½" in an hour. A mark was made every 1½". Each mark was labeled, starting at 7 P.M. and continuing to

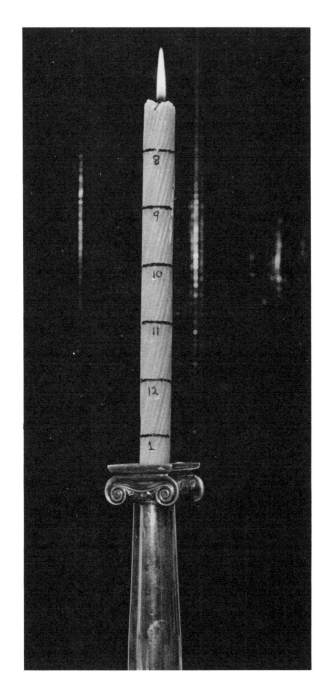

Candle Clock

1 A.M. A faster-burning candle might not burn that many hours, so just mark as many hours as the candle will accommodate.

You can start your candle clock at any hour you wish, day or night. Just remember that the first line that the candle burns down to will be one hour later than the time you lit it. To keep the candle burning at an even rate, place it away from drafts.

Candle clocks and water clocks are good things to have around when the electricity goes out, which happens in the country during bad storms. A candle clock should *never* be used in the bedroom at night. You might fall asleep with it still burning.

In dry countries another device was used whereby sand ran down from one container into another. The hourglass finally evolved from that. Every New Year's holiday Father Time is usually shown with his hourglass—a favorite symbol of the passing of time.

Even after the invention of mechanical pendulum clocks, hourglasses continued to be used at sea because pendulums couldn't work in a rolling and pitching ship. The "hour" glasses aboard ship were actually half-hour glasses. Sailors took turns standing "watch." In other words, they were responsible for steering the ship and adding or lowering sails if there was a change in the weather. Each "watch" was "stood" for a 4-hour period.

Do you know about the system of bells that has continued in use even with present-day ship's clocks? This system is based on the early half-hour glasses. When the first half-hour glass was turned over at twelve-thirty (timing started at twelve midnight), a bell was rung. Two bells meant one o'clock, 3 bells one-thirty, 4 bells two o'clock, and so on until the watch changed at 8 bells, or four o'clock. Now the tricky thing about this was that 8 bells could mean either four o'clock in the morning or in the afternoon, twelve midnight or noon. The only way they could tell which was which would be to look out a port to see if it was light or dark outside.

At any rate, the watch was changed (one sailor would replace the one who had been standing watch) every time 8 bells were rung and another 4-hour cycle began.

You still see "hour" glasses, but they are tiny and are now called "egg timers"; it takes 3 minutes for the sand to run down from one half of the glass into the other. If you have one, why not check it against a clock.

Chapter V

IS THERE A NOW?

BY NOW you have a pretty good idea how humans have solved the problems of measuring the passage of time. You have really gotten sort of a bird's-eye view of the history of mankind—at least from practically 10,000 years before the birth of Christ.

But what is time? Is it how the sun rises and sets? the way the stars seem to revolve around the earth? Or is it in the phases of the moon? No, all these are merely indicators of the passing of time. They have nothing to do with time itself.

Although they are certainly affected by it, monkeys, fish, and even racehorses do not care about time. Only people care about time. Time is only a figment of mankind's imagination. It has come from the brains of hundreds of people over the span of centuries.

However, in the beginning the people who worked out some of the basic ideas of time kept the ideas pretty much to themselves. They were shared with kings, emperors, bishops, and the like. But it wasn't until about 200 years ago that the ordinary, everyday person was allowed to tell time, with any accuracy, by himself.

The reason this precious information was withheld from ordinary people and reserved for the powerful was that it was a way of ensuring that they would continue to be in power. After all, supposing everyone had to come to your office to find out what time it was. Why, you'd have so much power you'd have to hire assistants to help you give the masses of people your time.

Even after the time was known by everybody, there were still problems to be solved. Perhaps an experience we had will give you a clue to one of the biggest problems people faced in measuring time.

Not too long ago we left Hong Kong on a jet at nine o'clock Sunday morning. We stopped over in Tokyo for half an hour and left there at twelve noon Hong Kong time. We were flying east toward the rising sun. But at four o'clock in the afternoon (Hong Kong time) the sun set. When it was ten o'clock "our time" that night, the stars and the blackness of night began to fade and dawn appeared outside our window. It was, of course, the dawn of a new day, but what day? Monday you say? Perhaps not. For what actually happened was that we took off from Hong Kong at nine o'clock Sunday morning, flew 20 hours, including stopovers in Tokyo and Seattle, and landed in New York at four o'clock Sunday afternoon (Eastern standard time).

The key to this mystery is the international date line. As you read this, it is tomorrow west of the international date line—unless, of course, you *are* west of the date line, in which case it's yesterday east of the date line. Confused? Well, let's see if we can confuse you some more.

Along the latitude, for example, sun time changes about one minute for every 12½ miles of travel east or west. In 1860 there were some 300 local time zones. On this basis, when it was noon in Chicago it was 12:31 in Pittsburgh, 12:24 in Cleveland, 12:13 in Cincinnati, 12:09 in Louisville, 11:50 in St. Louis, 11:48 in Dubuque, 11:41 in St. Paul, and 11:27 in Omaha. There were around 27 local times in Michigan, 38 in Wisconsin, 27 in Illinois, and 23 in Indiana.

The railroads were beginning to stretch across the country. With 300 local time zones, they found it impossible to write timetables. So before 1883 they started a system called "railroad time." That system reduced the number of time zones (for railroad purposes only) to 100. But this only made things more complicated. For example, the railroad station in Buffalo, New York, displayed 3 large clocks: one was set for local Buffalo time; another for New York City time, used by the New York Central Railroad; and a third showing Columbus, Ohio, time was used by the Michigan and Southern Railroad. In Pittsburgh there were 6 different kinds of time for confused and annoyed passengers to choose from. Obviously, there were many different "weres," "nows," and "thens." Even now time is an enormously complicated subject, but around 1880 it was ridiculous. Something had to be done.

It wasn't that people hadn't been suggesting solutions to the problem. Professor Charles Dowd recommended that time be changed by an hour in belts (running north to south) 15° wide. That would divide the United States into 4 time zones. This plan was submitted to the railroads in 1872 and was finally adopted 11 years later in 1883.

Because the fewest trains would be running on Sunday, the big shift from 100 railroad times (and, because the people were so impressed by this new system, from the 300 local times as well) to 4 standard time zones occurred at noon, Sunday, November 18, 1883. The exact time, provided by the United States Naval Observatory in Washington, D.C., was telegraphed to all parts of the country at the same moment. As Harrison J. Cowan points out in his book, *Time and Its Measurement,* "this became known as 'the day of the two noons.'" In the eastern part of each zone clocks had to be set back from 1 to 30 minutes. After they had passed "noon" according to their old local time, another "noon" was telegraphed from Washington, D.C.

When the new standard time went into effect, it was a major happening. Some people were afraid it was "against God's laws," insisting that "the railroads have no right to interfere with the movements of the sun." How one could "interfere with the movements of the sun" by resetting hundreds, thousands, or even millions of clocks remains a mystery. The idea of it being "against God's laws" is a strange idea in view of the fact that time is entirely a creation of the human mind.

However, most people, particularly those who did any traveling, were so tired of all the confusion about what time it was at a given spot that they were greatly relieved when it was simplified.

The United States was not the only country that faced a confusion of time zones. The problem was worldwide. The Prime Meridian Conference of 1884 established a universal solar mean time. What it accomplished was the division of the world into 24 longitudinal time zones, each 15° wide. The center of the first zone was marked 0° longitude and was on a line ("meridian") that ran through the Greenwich (England) Observatory. Twelve time zones, or 180° (12×15°) from Greenwich—in other words, as far as you can get from England or halfway round the world—were established as the international date line.

For each of the zones east of Greenwich one hour is added to Greenwich time. Greenwich is the standard against which all time is compared. Aviator's clocks are often set to Greenwich time. Pilots refer to it as "Zulu time." Aviators use a phonetic alphabet. "Alpha, Bravo, Charlie" are the first three code words in that alphabet. "Zulu" is the last and, in this case, "Z" stands for "zero degrees"—the location of the Greenwich Observatory. For each of the time zones west of Greenwich one hour is subtracted from Greenwich time. As you may already have figured out, sooner or later the minus time zones are going to bump into the plus time zones. The most logical place for them to meet is 180° away from Greenwich.

"Well," you might say, "I don't happen to live in England. What happens to time zones in the United States?" If you live in New York, you're in the Eastern time zone. In addition, almost everybody as far west as Cincinnati is in the Eastern time zone. The next time zone, going west, is the Central time zone. This zone includes Chicago, Minneapolis, and Omaha. The line jogs west to take in most of the state of Texas. The time zone that includes most of the Rockies is called, appropriately enough, the Mountain time zone. This zone includes most of Idaho and Utah and all of Montana, Wyoming, Colorado, Arizona, and New Mexico. Parts of Idaho and Utah and all states west of that, including Washington, Oregon, Nevada, and California, are in the Pacific time zone.

Here's the way it works: If it's nine o'clock at night in the Eastern time zone, it's eight o'clock Central, seven o'clock Mountain, and six o'clock in the evening in the Pacific time zone. Let's see what happens when you go east from the Eastern time zone. Supposing it's nine o'clock Monday night and you take off on a jet for London. Now you're heading east and traveling at a little more than 500 miles per hour toward the dawn of a new day. At the very moment of your takeoff it's two o'clock Tuesday morning in London (Greenwich time). Supposing your trip takes you 6 hours and you haven't reset your watch. You'll land in London at three o'clock Tuesday morning (New York [Eastern] time). But the sun will be well up in the sky in London (if it isn't raining) when you arrive, and it will be eight o'clock Tuesday morning (Greenwich time).

So, you see, when "now" is depends on where you are. But the fact remains that no matter where you are, the time "now" cannot be measured with any known clock. Bear in mind that the minute hand on a clock moves slowly but continuously. You may say, "I have a digital clock and the minute number stays up for 60 seconds." But we can reply that if you stop the clock in your kitchen, time will seem to stand still until the clock starts again. But that only proves what we said in the beginning: Clocks don't "make" time, they only measure it—and with varying degrees of success at that.

By explaining all of this, we are just trying to make you think in new ways about time and the word "now." The truth is that "now" is a perfectly useful word. It refers to what is going on in the present. Here's a further complication about the idea of "now." By the time you say the word "now" it will already have become "then." Depending on what you're talking about, the present can mean the minute you're speaking or what you've done in the past hour. Or, if you're reading a book on history, the author may consider the past 10 years as the present. An archeologist or geologist may consider the past 2,000 years as the present.

All we are saying is that there are no scientific definitions for words like "then" or "now." And yet they are both words and ideas that you and everyone else use hundreds of times a day. People understand what is meant by the words even if they can't be defined scientifically. It is much more important to make yourself understood by other people than to know precisely what time it is.

APPENDIX

PLACE	LATITUDE (N)	LONGITUDE (W)	CORRECTION (MINUTES) SUN TIME TO STANDARD TIME
Atlanta, Ga.	34°	84°	+36
Birmingham, Ala.	34°	87°	−12
Boise, Ida.	44°	116°	−16
Boston, Mass.	42°	71°	−16
Buffalo, N.Y.	43°	79°	+16
Butte, Mont.	46°	113°	+32
Carson City, Nev.	39°	120°	0
Charleston, S.C.	33°	80°	+20
Chicago, Ill.	42°	88°	−8
Cincinnati, Ohio	39°	84°	−24
Cleveland, Ohio	41°	82°	+28
Dallas, Tex.	33°	97°	+28
Denver, Colo.	40°	105°	0
Des Moines, Iowa	42°	94°	+16
Detroit, Mich.	42°	83°	−28
Fargo, N.D.	47°	97°	+28
Indianapolis, Ind.	40°	86°	−16

PLACE	LATITUDE (N)	LONGITUDE (W)	CORRECTION (MINUTES) SUN TIME TO STANDARD TIME
Jackson, Miss.	32°	90°	0
Jacksonville, Fla.	30°	82°	+28
Kansas City, Mo.	39°	95°	+20
Little Rock, Ark.	35°	92°	+8
Los Angeles, Cal.	34°	118°	−8
Louisville, Ky.	38°	86°	−16
Milwaukee, Wisc.	43°	88°	−8
Minneapolis, Minn.	45°	93°	+12
Nashville, Tenn.	36°	87°	−12
New Orleans, La.	30°	90°	0
New York, N.Y.	40°	74°	−4
Norfolk, Va.	37°	76°	+4
Omaha, Nebr.	41°	96°	+24
Philadelphia, Pa.	40°	75°	0
Phoenix, Ariz.	33°	112°	+28
Pittsburgh, Pa.	40°	80°	+20
Portland, Maine	44°	70°	−20
Portland, Ore.	45°	123°	+12
St. Louis, Mo.	39°	90°	0
Salt Lake City, Utah	41°	112°	+28
San Francisco, Cal.	38°	122°	+8
Seattle, Wash.	48°	122°	+8
Syracuse, N.Y.	43°	76°	+4
Washington, D.C.	39°	77°	+8
Wichita, Kan.	38°	97°	+28

BIBLIOGRAPHY

Sundials and Roses of Yesterday, by Alice Morse Earle. Macmillan, 1902; rpt. Chas. N. Tuttle, 1971.

Ye Sundial Booke, by T. Geoffrey and W. Henslow. W. G. Foyle, 1935.

Time and Its Measurement from the Stone Age to the Nuclear Age, by Harrison J. Cowan. World Publishing Co., 1958.

The Riddle of Time, by Thelma Harrison Bell and Corydon Bell. Illus. by Corydon Bell. Viking, 1963.

The 365 Days—The Story of our Calendar, by Keith G. Irwin. Thomas Y. Crowell, 1963.

Man and Time, by J. B. Priestly. Doubleday & Co., 1964.

Dictator Clock: 5,000 Years of Telling Time, by Roger Burlingame. Macmillan, 1966.

Time (Life Sciences Library Series), by Samuel H. Goudsmit, Robert Claiborne et al. Time-Life Books, 1966; rev. 1969.

The Nature of Time, by G. J. Whitrow. Holt, Rinehart & Winston, 1972.

Sundials—Their Theory and Construction, by Albert E. Waugh. Dover, 1973.

A Choice of Sundials, by Winthrop W. Dolan. Stephen Greene Press, 1975.

Henry Humphrey and Deirdre O'Meara-Humphrey are each authors of their own books. *When Is Now?* is the first book they have created together.

HENRY HUMPHREY is well known for his previous books, including *What Is it For?*, *What's Inside?*, *Sights and Sounds of Flying, The Farm,* and *Sailing the High Seas.* These books, covering a variety of interesting subjects, have been enjoyed by countless young readers for their clear, readable style and striking photographs.

DEIRDRE O'MEARA-HUMPHREY is a fine craftsperson with the unique ability to make other people understand how to make the objects themselves—a skill demonstrated in such books as *Plastic Crafts* and *Creating with Plastics.* Her clear, easily grasped instructions and her precise line drawings are perfect teaching tools for any beginning enthusiast.